EVERYONE EATS™

RICE

Jillian Powell

RAINTREE STECK-VAUGHN
PUBLISHERS
The Steck-Vaughn Company

Austin, Texas

Titles in the series

BREAD EGGS FISH FRUIT MILK
PASTA POTATOES RICE

Published by Raintree Steck-Vaughn Publishers,
an imprint of Steck-Vaughn Company
Everyone Eats™ is a trademark of Steck-Vaughn Company.

Library of Congress Cataloging-in-Publication Data
Powell, Jillian.
Rice / Jillian Powell.
p.   cm.—(Everyone Eats)
Includes bibliographical references and index.
Summary: Provides information on the cultivation, consumption,
nutritional value, and varieties of rice, as well as several recipes.
ISBN 0-8172-4758-0
1. Rice—Juvenile literature.
2. Cookery (Rice)—Juvenile literature.
[1. Rice.]
I. Title. II. Series: Powell, Jillian. Everyone Eats.
TX558.R5P69   1997
641.3'318—dc20   96-32831

Printed in Italy. Bound in the United States.
1 2 3 4 5 6 7 8 9 0 01 00 99 98 97

Picture acknowledgments
Cephas title page, 21 (bottom), 24 (top); Chapel Studios 24 (bottom); Bruce
Coleman 10, 11, 12 (top), 15 (top), 16, 17 (right), 23 (top); Eye Ubiquitous contents
page, 5, 8, 12 (bottom), 15 (bottom), 18, 20; Holt Studios International
17 (left); Hulton Deutsch 6 (both), 7; Hutchison Library 4, 13; Life File 14, 19
(both), 22, 23 (bottom); Wayland Picture Library 9 (both), 21 (top), 25.

Contents

Remarkable Rice **4**

Rice in the Past **6**

The Food in Rice **8**

What Is Rice? **10**

How Rice Grows **12**

Harvesting Rice **14**

From Farm to Table **16**

Ways of Using Rice **18**

Cooking Rice **20**

Customs and Beliefs **22**

Rice Dishes from Around the World **24**

Rice Recipes for You to Try **26**

Glossary **30**

Books to Read **31**

Index **32**

Remarkable Rice

Rice feeds over half of the world's population every day. After wheat, rice is the second most widely grown crop in the world. It is one of the oldest foods and has been grown and eaten in a similar way for thousands of years.

More than one billion people spend their lives growing rice, which is eaten in almost every country in the world. Most rice is eaten in Asia, where it is cooked for breakfast, lunch, and dinner. Asian people each eat up to 330 pounds of rice every year.

▲ A Chinese family enjoying their evening meal. Almost every meal in China is served with rice, and everyone in the family has his or her own rice bowl.

There are more than 7,000 different kinds of rice. The rice grains are different in size, shape, color, taste, and smell. The best-known types are long grain, short grain, and brown rice, but there is also pink, blue, yellow, purple, and even striped rice! There are now rice banks that store grains of every kind of rice.

In China, where about one third of all rice grows, rice is so important that people sometimes greet each other by saying "Have you eaten rice today?" In languages like Japanese and Indian Sanskrit, the word for "food" also means "rice." On the island of Madagascar, off the southeast coast of Africa, rice is so important that it was once used as money. Even today it is still used to measure time, so half an hour is "the time it takes to cook rice."

▶ In many countries children help their parents with the rice crop. This girl from Sri Lanka is planting rice seedlings.

In Asian countries, there are some birds, such as the Java sparrow, that live and feed in the rice fields. They are known as "rice birds."

Rice in the Past

People were collecting and planting wild rice seeds in parts of Asia as long as 7,000 years ago. The practice of growing rice spread to Africa and from there to Europe through traders, armies, and wandering tribes. From Roman times onward, rice was taken from Africa to Europe on ships carrying valuable spices. In Great Britain in the Middle Ages rice was thought of as rare and valuable, and it was stored in locked cupboards with expensive spices. People ate rice in sweet puddings, and it was often fed to babies and sick people.

In the fifteenth century, ships like this carried rice and spices to Europe.

The Italian explorer Marco Polo is said to have visited China in the thirteenth century and found the Chinese growing many kinds of rice, including types with pink, white, or yellow grains. Rice was served in separate bowls so the special flavors could be tasted. It was also made into flour and wine.

The thirteenth-century Italian explorer Marco Polo

In the seventeenth century, explorers and settlers carried rice grains to the New World. There is a story that a ship from Madagascar was on its way to England but was blown off course and landed near the Carolinas for repairs. The captain left behind a sack of rice grain, which is how the American rice trade began.

In Great Britain, rice continued to be used mainly for sweet puddings. In the seventeenth century, cooks made rice pudding with nutmeg, mace, rosewater, sugar, and eggs. By the eighteenth century, there was a recipe for an "economical rice pudding," made by tying rice and raisins in a cloth and boiling them.

To the Ancient Greeks and Romans, rice was rare. It was sometimes used like a medicine to help cure the sick.

▼ This nineteenth-century Chinese picture shows that rice has been planted in the same way for centuries.

The Food in Rice

Rice is a healthy food. Rice grains contain starch, a kind of carbohydrate. When a rice seed is planted, the starch it contains provides food for the new plant. This starch is also an important food for us, because we need carbohydrates to give us energy. About six ounces of rice gives us enough energy to walk for forty-five minutes, to dance for twenty minutes, or to participate in sports for fifteen minutes.

Food scientists discovered the importance of vitamins about 100 years ago. Some chickens being fed on a diet of white rice became ill, but when they were put on a diet of brown rice, they recovered. Scientists discovered the reason—the important B vitamins that bran contains.

▼ Wedding guests in southern India eating rice served on banana leaves

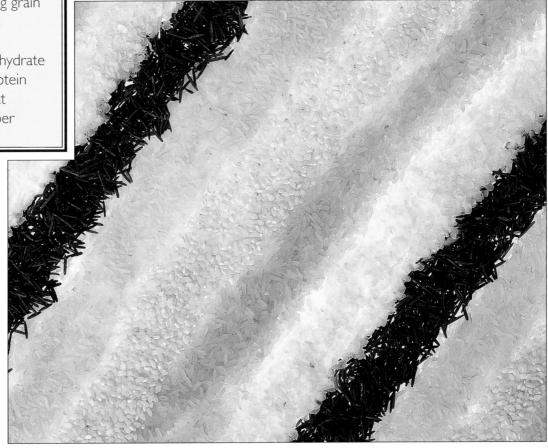

Bran in the outer layer of rice grains contains water, protein, some minerals, and B vitamins. Vitamins and minerals help keep our bodies healthy. Bran also provides fiber, which helps us to digest and pass food through our bodies. Brown or whole-grain rice still has some of its bran layer, so it is more nutritious than white rice, which has had the bran removed.

▲ A selection of uncooked rice. The black grains are wild rice.

▼ For a balanced meal, rice can be served with foods like eggs, fish, vegetables, or meat.

Rice can lose some of its vitamins into the water when it is boiled. Chinese people cook rice by steaming it, which helps to keep in more of the vitamins. Sometimes, a coat of vitamins is added to rice grains to replace some of the nutrients, but these can still be lost during cooking.

What Is Rice?

Rice is a type of grass. It is a cereal crop, like wheat, oats, barley, and rye. The rice we eat is the seed of rice plants. We eat many kinds of seeds, such as beans, peas, and lentils. Rice grows in many different places, from high mountains to river valleys and mangrove swamps by the sea. Some rice (called upland rice) can grow on dry land but most rice grows in flooded fields called paddies. "Paddy" is a Malaysian word meaning "rice growing in deep water," and it is also used to describe raw rice grains.

▼ Rice plants being harvested from a typical rice paddy in West Sumatra, Indonesia

▶ The head of a rice plant. A head of rice has many spikes, each carrying rice grains.

Different types of rice plants grow to between three and twenty feet high. They grow fastest in warm, wet lands, shooting up as much as ten inches a day when the water level in the paddies is rising. The plants grow as fast as they need to in order to keep their top leaves above water. Even if monsoon rains uproot a plant, special roots higher up its stem help it to stay alive until it can take root again.

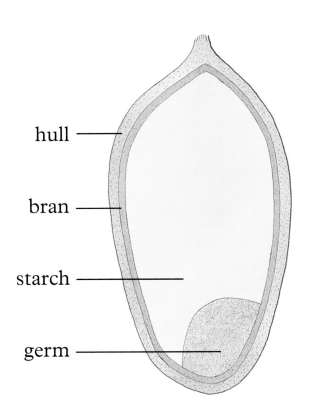

hull

bran

starch

germ

Rice plants take nutrients from the water through their hollow stems. They take in air through their leaves, stalks, and even underwater, through their roots.

The stalks produce long, flat leaves and heads that send out lots of little spikes. These carry the rice grains. Each grain has a tough, outer husk and a skin of bran that protects the starch and the germ inside, from which a new rice plant could grow.

11

How Rice Grows

Rice grows in many different lands and climates, but it grows best in the warm, wet countries of Asia. The countries of the "rice bowl" of Asia produce 90 percent of the world's rice. It is also grown in the United States, Africa, Australia, and Europe. Most of the world's rice grows in specially flooded paddy fields. Farmers build low mud walls around the fields and dig channels to carry water to the paddies. In mountain areas, terraces are cut into the hillsides to keep the soil from being washed away. The terraces are then flooded with water pumped from storage tanks. Water from rivers and monsoon rains provides the rice plants with food.

▲ Rice terraces cut into the hillside in Indonesia. The terraces keep soil from being washed down the hill.

▲ Oxen plow a paddy on the Indonesian island of Bali.

Before flooding the paddy, the farmer plows the earth using buffalo, oxen, or a tractor. The rice seeds are soaked in water overnight until they sprout. Then they are planted in nursery beds, where they grow for about thirty days before being planted in the paddy.

On rice farms in the United States and Australia, the rice fields are usually prepared by huge machines guided by computers. Tractors plow the fields and diggers dig water channels where pipes are laid. Rice grains can be sown by machines or from airplanes.

The rice crop is easily damaged by drought or hurricanes. Weeds can take water and food from the rice plants. Rats, worms, birds, snails, and fungus can damage the plants. Some farmers use chemicals to kill weeds, pests, and diseases that could harm the crop. In the United States and Australia, airplanes may be used to spray the chemicals.

▶ A man weeding rows of rice in the Philippines. Weeding is a tiring job.

Harvesting Rice

As rice ripens in the sunshine, it turns golden. Rice must be harvested at just the right time. If it is harvested too early, the grains will be moist and green and may rot. If the farmer leaves it too late, the grains will be too dry and may crack.

Harvesting is done by hand or machine. In parts of Asia, hand tools such as knives and scythes are used. In the huge rice fields of the United States and Australia, combine harvesters gather in the rice.

▼ A combine harvester gathers in the rice harvest in Spain.

Once harvested, rice plants must be threshed to separate the grains from the rest of the plant. On big western rice farms, combine harvesters do this job. In Asia, it is often done by hand. Sheaves of rice are beaten against logs or slatted bamboo screens or are trampled by farm animals. Some farmers have small threshing machines powered by hand, foot, or gasoline engine.

People in Bali threshing rice by hand. As they beat sheaves of rice plants against a bamboo screen, the rice grains fall to the ground.

After threshing, the rice is winnowed to get rid of any chaff or bits of husk and straw. This can be done by a machine but, in poorer countries, winnowing is still done by hand and is traditionally a woman's job. The rice is tossed on a bamboo tray, so that the rice grains fall down onto a mat, and the chaff (husks) blows away on the wind. It is skilled work.

Two women from Sumatra winnowing rice in the traditional way

From Farm to Table

To prevent it from rotting, rice grain must be dried before it is stored. The grains may be raked out to dry in the sun or piled in sacks that have hot air blown through them. Rice must be stored away from damp, heat, rats, insects, birds, and fungi. If it is stored in a clean, dry place, rice will keep for several years.

In Pakistan, farmers store rice in large clay pots, and in parts of India goatskin bags are used. In many Asian countries, rice barns are built on stilts so that they keep dry even if the rivers flood. Farmers in Sri Lanka make rice storehouses from jungle creepers plastered with clay and lime and set on stilts. In western countries, rice is stored in huge steel or concrete silo towers, where the temperature and moisture of the air can be controlled.

▲ Children from Sumatra spread out the rice to dry in the sun.

16

A basket used for storing rice in the roof of a hut in Guinea, West Africa

At this stage, the rice still has its husk and a layer of bran. Some rice is milled to remove the husk and bran, leaving what we call white rice. In Asian countries, milling may be done by hand, by pounding the grain in a wooden bowl using a long pole. About one quarter of the world's rice is milled in modern rice mills where the grain is put into metal drums and rollers remove the husks. Modern machines sort grains into size, type, and color for packaging. White rice may be polished with a glossy coat of glucose (sugar) and talcum and is popular in the United States and parts of Europe.

Milling rice by hand in Thailand

17

Ways of Using Rice

After it has been milled, rice can be used in lots of different ways before being sold. There are products like "boil in the bag" rice, instant rice, flaked rice, and rice pudding in plastic cups. Rice-based breakfast cereals are made from rice that has been steamed and toasted, and then coated with sugar or chocolate. Some rice is cooked then frozen or dried ready for use in products like soup and baby food. Broken rice may be used in sausage fillings and canned pet food. It is also made into vinegar for cooking and rice wines and beers, such as Japanese sake and Malaysian and Chinese beers.

▼ Colorful Japanese candies made from rice

In the early twentieth century, an American biochemist discovered how to make puffed rice by firing grains through a gun.

Rice straw being transported in boats on China's Grand Canal. Rice straw is used to make paper.

Rice is also ground into flour to make cakes, noodles, and face powder. Starch from rice is used to make the rice paper for cakes like macaroons and to stiffen clothes and make glue.

No part of the rice plant is wasted. After harvesting, some farmers plow the stubble back into the soil to feed the next crop. Rice straw is woven into baskets, hats, mats, shoes, and ropes or used for thatching, animal food, or bedding. The husks, which are removed by milling, can be used to make light bricks and packing materials. More than 75 million tons of husks are produced every year. The bran, which contains lots of food value, can be used for animal food, fertilizer, or oils for cooking or oiling machinery.

 Japanese rice wine is traditionally served from china bottles and cups.

Cooking Rice

All rice is hard and gritty before it is cooked. Different types of grains cook in different ways. The long, thin grains of long grain rice stay separate and fluffy when cooked. The short, round grains of glutinous or sticky rice, such as arborio or Carolina rice, stick together. They are often used to make rice puddings or risottos.

Rice can be cooked by boiling it in water, wine, milk, or meat or vegetable stock. It can also be steamed over boiling water or fried in fat, then boiled.

The starch in rice makes it sticky. Some rice has to be rinsed in water before cooking, to get rid of the starch. If it were not rinsed, the rice grains would stick together during cooking.

▲ The Chinese often steam rice in bamboo baskets over boiling water. This is a healthy way to cook rice because it helps to keep in more protein and vitamins.

20

As it cooks, rice absorbs liquid, so it fluffs up to three times its size. For every cup of rice you need about one and a half to two cups of water. Boil the water, add the rinsed rice, and stir it in. Then turn the heat down so that the water is just simmering. The rice will take about twenty to twenty-five minutes to absorb all the water depending on the kind of rice you use.

▶ Brown rice takes longer to cook than white rice and is more chewy.

In China, southeast Asia, India, and parts of Europe, rice is sometimes fried and colored with spices like saffron and turmeric. It is served with spicy dishes, such as chili con carne and curries, or cooked and eaten cold with salads.

◀ Spices have been used to add color and flavor to this rice salad dish.

Customs and Beliefs

For thousands of years, rice has been an important part of religious life in Asia. Many ancient peoples worshiped the rice goddess, offering her gifts at planting and harvest times to make sure that there would be plenty of grain for the next crop. On the Indonesian island of Bali, small shrines are still built in the paddies, and rice cakes are offered to the rain gods to bring rain and a good harvest. In Japan, there is an ancient tradition that the Emperor begins the rice planting ceremony. In Sri Lanka, farmers talk to astrologers and earth magicians before planting the rice. If the crops fail, the men from the village must calm the rice goddess by fighting each other with coconuts.

In India, there is a special thanksgiving ceremony, called Pongal, for the rice harvest. At other important festivals, people decorate their doorsteps with Rangoli patterns, made from colored rice flour, to welcome visitors.

▲ In Japan, rice is made into lucky charms to bring a good harvest.

▲ Farmers in Bali make offerings to the rain gods, hoping for good weather.

Rice is often eaten at celebrations. The Chinese make a dish for special occasions called Eight Treasure Rice pudding, which includes dates, candied fruits, and nuts. The Japanese traditionally eat rice noodles for birthdays and at the New Year, as they are believed to bring luck and a long life. Rice wine is drunk on special days like the Festival of the Moon in the autumn. At Hindu, Muslim, and Christian weddings, it is traditional to throw rice at the bride and groom to bring luck and children to the marriage.

▶ Rice is thrown at the bride and groom at a Christian wedding in France.

Rice Dishes from Around the World

Rice is eaten in thousands of different ways around the world.

It can be served boiled, steamed, or fried. In China, Japan, and other eastern countries, rice is eaten from a bowl, using chopsticks. In parts of India, people roll it into balls and eat it with their fingers.

The Chinese eat a kind of rice porridge, called congee, for breakfast. It can be mixed with bean curd, pickles, and chicken or other meats. In southern China, rice noodles are very popular, especially when served with seafood and in soups and stir-fry dishes.

▲ In Japan, rice is used to fill parcels of raw fish called sushi.

In India, there are many popular savory rice dishes, such as pilau rice and biriani. Sweet rice dishes like zarda and khir are also eaten.

◄ A sweet Indian rice pudding called khir, made with almonds and raisins and flavored with rosewater, saffron, and cardamom seeds. Hindus often give khir to their priests on religious occasions.

24

In West Africa, jollof rice is eaten on special occasions. Different ingredients are used in different regions and include beef or chicken, onions, chilis, tomatoes, and vegetables. South African cape kedgeree is rice cooked with flaked fish, egg whites, milk, and curry powder, with hardboiled eggs sliced on top.

▲ The Spanish use rice for their national dish, paella, which was once made with snails, eels, and green beans but now usually contains tomatoes, peas, and seafood such as prawns and mussels.

In Europe, Italian risotto is traditionally made from rice cooked with onions, mushrooms, and peppers in wine. The Greek dish dolmades is rice wrapped in vine leaves.

In the southern United States, the traditional dish jambalaya is made from rice cooked in a stew with shrimps, chicken, and turkey.

The traditional Jamaican dish called "rice and peas" is, in fact, usually made with rice and beans.

Rice Recipes for You to Try

Chinese Fried Rice

To serve four people you will need:

1-1/3 cups long grain rice, cooked according
    to the instructions on the package
2 slices of bacon, chopped into strips
1/3 of a 10-ounce package of frozen peas
2 tablespoons cooking oil
1 teaspoon salt
2 eggs, beaten
4 ounces bean sprouts
2 tablespoons chopped scallions

1 Ask an adult to help you heat about 2 cups of water in a saucepan until it is boiling. Add the peas and bring the water back to the boil.
Turn down the heat and let the water bubble fast without boiling over for about two minutes. Drain the peas.

2 Ask an adult to help you to heat the oil in a wok or large frying pan until it is hot. Add the cooked rice and stir with a wooden spoon to separate the grains.

3 Cook for about one minute, then add the bacon, peas, and salt. Fry for about five minutes on a high heat.

4 Now add the beaten eggs and bean sprouts and fry for another two minutes, until the eggs have set.

5 Spoon the rice mixture onto a hot plate and sprinkle with chopped scallions.

Caribbean-style Rice

To serve four people you will need:

1-1/3 cups long grain rice (check on the
    package whether you need to rinse
    the rice before use)
4 cups water
1-1/2 tablespoons butter
1 onion
2 ounces ham, chopped
1 red pepper
1 tablespoon cheese, grated
handful of chopped parsley
2 teaspoons salt

1 Ask an adult to help you heat the water in a pan. When it is boiling, add the salt then carefully pour in the rice. Turn down the heat so that the water is just simmering. Cover the pan and let the rice cook for about twenty minutes until it has absorbed almost all the water.

2 When the rice is almost cooked, add the butter, ham, chopped onion, and chopped pepper. Stir all the ingredients together, then cover the pan again and cook gently for another ten minutes.

3 Remove the rice from the heat and serve in a hot dish, sprinkled with the parsley and grated cheese.

Glossary

astrologers People who study the stars and planets in the belief that they affect our lives.

bean curd A food made from crushed soya beans.

biochemist Someone who studies the chemistry of living organisms.

bran The husk or outer coat of a cereal grain.

brown rice Rice that still has its layer of bran.

calories Measurements of the energy in food.

carbohydrate A type of food, including sugars and starches, that gives us energy.

cereal Any plant that produces grain used for food.

combine harvester A machine that combines two jobs, harvesting crops and separating the grain from the rest of the plant.

drought Shortage of water due to lack of rain.

earth magician Someone who claims to tell the future by examining lines and patterns in the earth.

fiber The part of food that helps us to digest food and pass it through our bodies.

fungus A kind of plant. Some can harm other plants or animals.

germ The part of a seed from which a new plant may grow.

husk The outer coat of a seed.

mangrove swamp An area of tropical trees and shrubs at the mouth of a river or by the coast that is flooded by the sea every day.

Middle Ages A period of history in Europe from about A.D. 500 to 1500.

milled Ground up. Seeds and cereal grains are often milled.

minerals Substances found in food that we need to keep our bodies healthy.

monsoon Winds that bring heavy rain at a certain time of year.

New World The Western Hemisphere.

nursery beds Areas where young plants are grown.

nutrient Anything in food that is needed for health, such as vitaimins and protein.

nutritious Containing nutrients.

paddy field A field that has been flooded for growing rice.

protein A substance found in certain foods. We need protein to grow and repair our bodies.

scythe A long curved knife used for cutting grass or crops.

seed The part of a plant from which a new plant may grow.

sheaves Bundles.

shrines Temples.

silo A building for storing grain.

staple diet The main or most important foodstuff in a person's diet.

starch A type of carbohydrate.

stir-fry A method of frying food quickly in oil, stirring all the time.

stock Water flavored with vegetables, meat, or bones.

talcum A fine powder made from soapstone.

terraces Flat areas of land cut into a hillside.

threshing Separating the grain from the rest of a plant.

upland rice Rice that is grown on high land.

vitamins Substances found in food that we need to keep us healthy.

white rice Rice that has been polished after the bran has been removed.

whole grain A cereal grain from which the bran has not been removed.

winnowing Separating chaff (bits of husk and stalk) from cereal grains.

Books to read

Food Around the World series. New York: Thomson Learning. A series of ten books on food by country. Each book presents the particular ways in which the peoples of different countries prepare and serve their food. Simple recipes are also provided.

Dibble, Lisa. *Food and Farming*. New York: Dorling Kindersley, 1993.

Johnson, Sylvia. *Rice*. New York: Lerner Books, 1985.

Moss, Miriam. *Eat Well*. New York: Macmillan Children's Group, 1993.

Perl, Lila. *Junk Food, Fast Food, Health Food: What America Eats and Why*. New York: Houghton Mifflin, 1980.

Tames, Richard. *Food: Feasts, Cooks, and Kitchens*. New York: Franklin Watts, 1994.

Index

Numbers in **bold** show subjects which appear in pictures.

Africa 5, 6, 11
Ancient Greeks 7
Ancient Romans 7
arborio rice 20
Asia 4, 6, 12, 14, 16, 17, 21, 22
Australia 11, 13, 14

Bali **12**, **15**, 22, **23**,
beer 18
birds 5, 13, 16,
biriani 24
bran 9, 11, 17, 19
breakfast cereals 18

calories 8
cape kedgeree 25
carbohydrate 8
Carolina rice 20
chili con carne 21
China **4**, 5, 6, 7, **8**, 9, 18, **19**, 21, 23, 24
combine harvesters **14**, 14
congee 24
curry 21

dolmades 25
drought 13
drying rice **16**, 16

Europe 6, 11, 17, 21, 25

fiber 9
France **23**

Great Britain 7
Greece 25
Guinea **17**

harvesting **10**, **14**, 14
Hindus 23
hurricanes 13

India **8**, 16, 21, 22
Indonesia **10**, **12**, 22

Jamaica 25
jambalaya 25
Japan 5, 18, **22**, 22, 23, 24
jollof rice 25

khir **24**, 24

Madagascar 5, 7
Malaysia 10, 18
Middle Ages 6
milling **17**, 17, 18, 19
minerals 9
monsoon 11, 12

New World 7

paella **13**
paddy fields **5**, **7**, **10**, 10, 11, **12**, 12, **13**, **14**, 22
Pakistan 16
Philippines **13**
pilau rice 24
plowing 12, 13
Polo, Marco **6**, 6
Pongal 22
protein 19

rain gods 22
Rangoli 22
rats 13, 16
recipes
    Chinese fried rice 26-7
    Caribbean style rice 28-9
rice goddess 22
rice paper 19
rice pudding 7, 18, 20
risotto 20, 25

South Africa 25
Spain **14**
spices 6, **21**, 21
Sri Lanka **5**, 16, 22
starch 8, 11, 19
storing rice 16, **17**
Sumatra **15**, **16**
sushi **24**

terraces **12**, 12
Thailand **17**
threshing 14, **15**

United States 7, 12, 13, 14, 17

vitamins 9

weddings **8**, **23**, 23
weeding **13**, 13
West Africa **17**, 25
West Sumatra 10
wine 18, **19**
winnowing **15**, 15

zarda 24